Synthesis Lectures on Ocean Systems Engineering

Series Editor

Nikolas Xiros, University of New Orleans, New Orleans, LA, USA

The series publishes short books on state-of-the-art research and applications in related and interdependent areas of design, construction, maintenance and operation of marine vessels and structures as well as ocean and oceanic engineering.

Alexander Arnfinn Olsen

Introduction to Ship-to-Ship Transfers for Oil Tankers

Alexander Arnfinn Olsen ⓘ
Southampton, UK

ISSN 2692-4420   ISSN 2692-4471 (electronic)
Synthesis Lectures on Ocean Systems Engineering
ISBN 978-3-031-74801-1   ISBN 978-3-031-74802-8 (eBook)
https://doi.org/10.1007/978-3-031-74802-8

This Springer imprint is published by the registered company Springer Nature Switzerland AG
The registered company address is: Gewerbestrasse 11, 6330 Cham, Switzerland

If disposing of this product, please recycle the paper.

Preface

Ship-to-Ship transfers for oil tankers involve the transfer of oil or petroleum products between two oil tankers while they are floating side by side at sea. This method is used for various reasons such as transferring oil cargo between larger and smaller tankers, transferring oil in areas where port facilities are limited, or transferring oil between tankers to optimise shipping routes. It is a common practice in the oil and gas industry to efficiently move oil from one vessel to another without the need to dock at a port. Ship-to-Ship transfers for oil tankers are typically carried out using specialised equipment and following strict safety protocols. Essentially, the Ship-to-Ship transfer process is usually conducted by:

- *Preparation*: both oil tankers involved in the transfer need to be properly positioned next to each other at a safe distance. Crew members from both vessels will prepare for the transfer operation
- *Mooring*: mooring lines and fenders are used to secure the two tankers together to ensure they remain stable during the transfer process
- *Hoses and equipment setup*: specialised hoses and equipment are used to connect the oil transfer systems of both tankers. These hoses are often equipped with emergency release mechanisms for safety
- *Pumping*: oil is pumped from one tanker to the other using high-capacity pumps. Flow rates and pressures are carefully monitored to prevent spills or accidents
- *Monitoring and safety*: throughout the transfer process, crew members on both tankers closely monitor the operation for any signs of leakage, overflows, or other potential hazards. Contingency plans are in place in case of emergencies; and
- *Completion and disconnection*: once the transfer is complete, the hoses are disconnected, and equipment properly stowed. Both tankers then resume their voyages.

It is important to note that ship-to-ship transfers for oil tankers require careful planning, coordination, and adherence to safety regulations to mitigate the risks associated with handling oil at sea.

This book provides an outline of the Ship-to-Ship transfer process, and how ship's masters and officers can conduct this inherently dangerous activity safely.

Southampton, UK   Alexander Arnfinn Olsen

Acknowledgements It is with the deepest gratitude that I thank the team at Babcock Marine and Technology for sharing their knowledge and insights during the development of this guide. I would also like to extend my thanks to Dr. Dieter Merkle and the editorial team at Springer for their assistance in bringing this guide together; and last but certainly not least, my deepest thanks go to my wife, Fidaa, who as always, has stood as steadfast as an anchor in the most violent of seas.

To you all, my sincerest thanks and gratitude.

The original version of the book has been revised. A correction to this book can be found at https://doi.org/10.1007/978-3-031-74802-8_10

Contents

Abbreviations and Acronyms

AIS	Automatic Identification Systems
cm	Centimetre
COLREGS	International Regulations for Preventing Collisions at Sea, 1974
EEZ	Exclusive Economic Zone
ETA	Estimated Time to Arrival
ft	Foot/feet
GHz	Gigahertz
H_2S	Hydrogen Sulphide
IGS	Inert Gas System
IMO	International Maritime Organisation
in	Inch
IOPCC	International Oil Pollution Prevention Certificate
ISGOTT	International Safety Guide for Oil Tankers and Terminals
ISM	International Safety Manual
ISPS	International Ship and Port Facility Security Code
MHz	Megahertz
MSDS	Material Safety Data Sheet
OCIMF	Oil Companies International Marine Forum
ORB	Oil Record Book
PDA	Portable Data Assistant
POAC	Person in Overall Advisory Control
SMS	Safety Management System
SOPEP	Shipboard Oil Pollution Emergency Plan
SS	Service Ship
STBL	Ship to be lightened
STCW	International Convention on Standards and Training, Certification and Watch Keeping Code
STS	Ship-to-Ship

SWL   Safe Working Load
UHF   Ultra-High Frequency
VHF   Very High Frequency
VLCC   Very Large Crude Carrier
VRP   Vessel Response Plan

List of Figures

List of Tables

Introducing the MARPOL Regulations

1

1.1 Introduction

The International Convention for the Prevention of Pollution from Ships (MARPOL) is arguably one of the most significant regulatory frameworks governing the maritime industry. Enforced by the International Maritime Organisation (IMO), MARPOL sets out strict guidelines and regulations aimed at preventing marine pollution, particularly from oil tankers. Exploring how MARPOL applies to oil tankers involves delving into the intricate details of the convention's provisions, such as discharge criteria, monitoring requirements, and compliance measures. Understanding the complexities of MARPOL in relation to oil tankers is crucial for ensuring environmentally sustainable practices within the shipping industry. By examining the nuances of MARPOL's application to oil tankers, this chapter seeks to illustrate the challenges and opportunities for improving pollution prevention efforts in maritime transportation.

1.1.1 Background on the International Convention for the Prevention of Pollution from Ships (MARPOL)

Since its inception in 1973, the International Convention for the Prevention of Pollution from Ships, known as MARPOL, has been a crucial framework in addressing the environmental impact of shipping activities. The convention was developed by the IMO in response to growing concerns about marine pollution caused by operational discharges and accidental spills from ships. MARPOL consists of six annexes, each focusing on specific aspects of pollution prevention, including oil, noxious liquid substances, harmful substances in packaged form, sewage, garbage, and air emissions. These annexes set out

A. A. Olsen, *Introduction to Ship-to-Ship Transfers for Oil Tankers*, Synthesis Lectures on Ocean Systems Engineering, https://doi.org/10.1007/978-3-031-74802-8_1

regulations and requirements for the design, construction, and operation of ships to min-imise environmental pollution. The establishment of MARPOL marked a significant step towards promoting cleaner and safer shipping practices on a global scale, highlighting the need for international cooperation in safeguarding the marine environment.

1.1.2   MARPOL Regulations for Oil Tankers

Oil tankers are subject to stringent regulations under MARPOL to prevent marine pol-lution caused by oil spills. These regulations are designed to reduce the environmental impact of transporting oil by sea and ensure that operators comply with best practices to safeguard the marine environment. MARPOL sets out specific requirements for oil tankers, including mandatory design and equipment standards to prevent oil spills and procedures for the safe transfer of oil. Compliance with MARPOL regulations is crucial to uphold the sustainability of marine ecosystems and protect vulnerable marine species from the harmful effects of oil pollution. By adhering to these regulations, oil tanker operators demonstrate their commitment to responsible environmental stewardship and contribute to the preservation of our oceans. Enforcement of MARPOL regulations for oil tankers is essential to promote sustainable shipping practices and mitigate the risks associated with oil transportation at sea.

1.1.3   Overview of MARPOL Annex I—Regulations for the Prevention of Pollution by Oil

These regulations aim to prevent pollution of the marine environment by oil through the implementation of various measures. The Annex I of MARPOL requires oil tankers to have a system that separates oil from water before it is discharged overboard. Fur-thermore, it sets specific requirements for the design and construction of oil tankers to prevent oil leaks and spills. Tankers are also required to have an oil discharge monitoring and control system, as well as an oil filtering equipment to ensure that any discharged oil meets the set standards. Inspections, surveys, and reporting requirements are integral to ensure compliance with the regulations, and violators may face severe penalties. Over-all, MARPOL Annex I is instrumental in safeguarding the marine environment from the harmful effects of oil pollution, focusing on prevention and enforcement to minimise the impact of oil tanker operations on the ecosystem.

1.1.4 Requirements for Oil Tankers Regarding Oil Pollution Prevention

To prevent oil pollution caused by oil tankers, the IMO has established stringent require-ments under MARPOL Annex I. Oil tankers are required to comply with a range of preventive measures, including having a double hull construction, which reduces the like-lihood of oil spills in case of a collision or grounding. Additionally, oil tankers must have segregated ballast tanks to prevent oil contamination during ballasting operations. The installation of high-capacity oily water separators onboard is mandatory to ensure that any oily water discharged overboard meets the set standards. Furthermore, oil tankers must have comprehensive oil spill response plans in place, which outline the steps to be taken in the event of an oil spill to minimise its impact on the marine environment. By adhering to these requirements, oil tankers can contribute to the prevention of oil pollution and help safeguard marine ecosystems.

1.1.5 Implementation of MARPOL Regulations on Oil Tankers

One major challenge in the implementation of MARPOL regulations on oil tankers is the issue of enforcement and compliance monitoring. While MARPOL sets out strict guide-lines and requirements for oil tankers to follow to prevent pollution and environmental harm, ensuring that these regulations are effectively enforced poses a significant hurdle. Many countries struggle with limited resources and expertise to effectively monitor and enforce compliance with MARPOL regulations. Furthermore, the international nature of the shipping industry complicates enforcement efforts, as different countries may have varying levels of commitment to enforcing MARPOL regulations. Additionally, some ship operators may attempt to circumvent regulations through illegal practices, such as illegal discharges of oil at sea. These challenges highlight the need for improved coordi-nation and cooperation among countries and the shipping industry to effectively enforce MARPOL regulations and ensure the protection of the marine environment.

1.1.6 Compliance Mechanisms for Oil Tankers Under MARPOL

One of the key compliance mechanisms for oil tankers under MARPOL is the Interna-tional Oil Pollution Prevention Certificate (IOPPC). This certificate is issued to vessels that meet the requirements of Annex I of MARPOL, which focuses on the prevention of pollution by oil. To obtain this certificate, oil tankers must undergo regular inspections by flag states or recognised organisations to ensure they comply with the regulations set out in MARPOL. Additionally, oil tankers must have Oil Record Books onboard to record all oil-related operations and disposals, which are subject to inspection during port state control inspections. These compliance mechanisms play a crucial role in ensuring that oil tankers operate in accordance with MARPOL regulations and help mitigate the environmental risks associated with oil transportation. By implementing these mechanisms effectively, MARPOL contributes to the global efforts to protect the marine environment from oil pollution.

1.1.7   Monitoring, Reporting, and Enforcement of MARPOL Regulations on Oil Tankers

One critical aspect of ensuring compliance with MARPOL regulations on oil tankers is the monitoring, reporting, and enforcement procedures put in place by regulatory bodies and flag states. Monitoring involves regular inspections of vessels to ensure they are adhering to MARPOL requirements, such as proper disposal of oily residues and ballast water management. Reports must be submitted by ships detailing their activities and any incidents that may occur, allowing authorities to track compliance and identify areas for improvement. Enforcement mechanisms, such as fines and penalties for non-compliance, serve as deterrents to ensure that operators adhere to regulations. Additionally, effective enforcement helps maintain a level playing field in the industry by holding all parties accountable to the same standards. Overall, the implementation of robust monitoring, reporting, and enforcement measures is crucial for upholding the integrity of MARPOL regulations on oil tankers and protecting the marine environment from pollution.

As noted, the MARPOL convention plays a crucial role in regulating the oil tanker industry and promoting environmental protection. Through its various annexes, MARPOL sets stringent standards for the prevention of marine pollution from oil tankers, including regulations for the design and equipment of vessels, as well as requirements for safe operations and contingency planning. Compliance with these regulations not only helps to minimise the environmental impact of oil spills but also contributes to the safety of crews and the sustainability of marine ecosystems. Additionally, MARPOL promotes international cooperation and coordination among member states to ensure consistent enforcement of regulations and the sharing of best practices. Overall, the implementation of MARPOL has significantly improved the environmental performance of oil tankers, leading to better protection of the marine environment and supporting sustainable shipping practices for the future.

Moreover, this chapter has delved into the various regulations and requirements imposed on oil tankers by MARPOL, highlighting the importance of preventing oil pollution and promoting sustainable practices in the shipping industry. This chapter has discussed the key provisions of MARPOL Annex I, which governs the prevention of pollution by oil, including strict limitations on oil discharges and the requirement for oil record books. Furthermore, this chapter has emphasised the need for oil tanker operators to comply with international standards and guidelines to ensure the safety of the environment and human health. In addition, the chapter has outlined the enforcement mechanisms of MARPOL, such as port state control inspections and the role of classification societies in verifying compliance with regulations. Overall, this chapter has provided a comprehensive overview of how MARPOL applies to oil tankers and the measures that must be taken to achieve environmental sustainability in the maritime sector.

1.1.8 Implications of MARPOL Regulations on Oil Tankers for the Maritime Industry

As MARPOL regulations continue to place stringent requirements on oil tankers, the maritime industry is facing significant implications. The implementation of MARPOL Annex I, which focuses on preventing pollution from oil spills, has led to the development and enforcement of strict standards for oil tankers regarding pollution prevention measures, such as double-hull design requirements and oil discharge restrictions. These regulations have increased operational costs for tanker owners, forcing them to invest in costly upgrades and advanced technologies to comply with MARPOL standards. Furthermore, the increased compliance requirements have also led to a shift in the industry towards more sustainable and environmentally-friendly practices. While these regulations aim to protect the marine environment from oil pollution, they have also brought about challenges for the maritime industry in terms of increased expenses and operational complexities. Overall, the implications of MARPOL regulations on oil tankers have prompted a necessary transformation in the maritime industry towards greater environmental responsibility and sustainability (Fig. 1.1).

1.2 Marpol Annex 1

Resolution MEPC.186(59) was adopted at MEPC 59 and contains the revised Chap. 8 to MARPOL Annex I on the prevention of pollution during the transfer of oil cargo between oil tankers at sea. These amended requirements are captured in Regulations 40, 41, and 42 from the new Chap. 8 to MARPOL Annex I and given below.

Start of Extract

1.2.1 Regulation 40 Scope of Application

(1) The regulations contained in this chapter apply to oil tankers of 150 gross tonnage and above engaged in the transfer of oil cargo between oil tankers at sea (STS operations) and their STS operations conducted on or after 1 April 2012. However, STS

Fig. 1.1 MARPOL
consolidated version 2022

operations conducted before that date but after the approval of the Administration of STS Operations Plan required under regulation 41.1 shall be in accordance with the STS Operations Plan as far as possible

(2) The regulations contained in this chapter shall not apply to oil transfer operations associated with fixed or floating platforms including drilling rigs; floating production, storage, and offloading facilities (FPSOs) used for the offshore production and storage of oil; and floating storage units (FSUs) used for the offshore storage of produced oil[1]

(3) The regulations contained in this chapter shall not apply to bunkering operations

(4) The regulations contained in this chapter shall not apply to STS operations necessary for the purpose of securing the safety of a ship or saving life at sea, or for combating specific pollution incidents in order to minimise the damage from pollution; and

(5) The regulations contained in this chapter shall not apply to STS operations where either of the ships involved is a warship, naval auxiliary or other ship owned or operated by a Flag State and used, for the time being, only on government non-commercial service. However, each Flag State shall ensure, by the adoption of appropriate measures not impairing operations or operational capabilities of such ships that the

STS operations are conducted in a manner consistent, so far as is reasonable and practicable, with this chapter.

1.2.2 Regulation 41 General Rules on Safety and Environmental Protection

(1) Any oil tanker involved in STS operations shall carry on board a Plan prescribing how to conduct STS operations (STS Operations Plan) not later than the date of the first annual, intermediate or renewal survey of the ship to be carried out on or after 1 January 2011. Each oil tanker's STS Operations Plan shall be approved by the Flag State Administration. The STS Operations Plan shall be written in the working language of the ship.

(2) The STS Operations Plan shall be developed taking into account the information contained in the best practice guidelines for STS operations identified by the International Maritime Organisation (IMO).[2] The STS Operations Plan may be incorporated into an existing Safety Management System (SMS) required by chapter IX of the *International Convention for the Safety of Life at Sea* (SOLAS), 1974, as amended, if that requirement is applicable to the oil tanker in question.

(3) Any oil tanker subject to this chapter and engaged in STS operations shall comply with its STS Operations Plan.

(4) The person in overall advisory control of STS operations shall be qualified to perform all relevant duties, taking into account the qualifications contained in the best practice guidelines for STS operations identified by the IMO.[3]

(5) Records[4] of STS operations shall be retained on board for three years and be readily available for inspection by a Party to the present Convention.

1.2.3 Regulation 42 Notification

(1) Each oil tanker subject to this chapter that plans STS operations within the territorial sea, or the exclusive economic zone of a Party to the present Convention shall notify that Party not less than 48 h in advance of the scheduled STS operations. Where, in an exceptional case, all of the information specified in paragraph 2 is not available not less than 48 h in advance, the oil tanker discharging the oil cargo shall notify the Party to the present Convention, not less than 48 h in advance that an STS operation will occur, and the information specified in paragraph 2 shall be provided to the Party at the earliest opportunity.

(2) The notification specified in paragraph 1 of this regulation[5] shall include at least the following:

(a) Name, flag, call sign, IMO Number and estimated time of arrival of the oil tankers involved in the STS operations

(b) Date, time, and geographical location at the commencement of the planned STS operations

(c) Whether STS operations are to be conducted at anchor or underway

(d) Oil type and quantity

(e) Planned duration of the STS operations

(f) Identification of STS operations service provider or person in overall advisory control and contact information; and

(g) A confirmation that the oil tanker has on board an STS Operations Plan meeting the requirements of regulation 41.

(3) If the estimated time of arrival of an oil tanker at the location or area for the STS operations changes by more than six hours, the master, owner, or agent of that oil tanker shall provide a revised estimated time of arrival to the Party to the present Convention specified in paragraph 1 of this regulation."

End of Extract

Notes

1. Revised Annex I of MARPOL, chapter 7 (resolution MEPC.117(52)) and UNCLOS article 56 are applicable and address these operations.
2. IMO's "Manual on Oil Pollution, Section I, Prevention" as amended, and the ICS and OCIMF "Ship-to-Ship Transfer Guide, Petroleum", fourth edition, 2005.
3. IMO's "Manual on Oil Pollution, Section I, Prevention" as amended, and the ICS and OCIMF "Ship-to-Ship Transfer Guide, Petroleum", fourth edition, 2005.
4. Revised Annex I of MARPOL Chaps. 3 and 4 (resolution MEPC.117(52)); requirements for recording bunkering and oil cargo transfer operations in the Oil Record Book, and any records required by the STS operations Plan.
5. The national operational contact point as listed in document MSC-MEPC.6/Circ.4 of 31 December 2007 or its subsequent amendments.

STS Transfer Operations Plan

2

2.1 Introduction

The STS Transfer Operations Plan (STS Plan) has been developed in accordance with the standards describe in MARPOL Annex I, as amended by Resolution MEPC.186(59), Chap. 8: Prevention of Pollution during Transfer of Oil Cargo between Oil Tankers at Sea, Regulations 40, 41, and 42. The STS Plan has been developed taking into account the information contained in the best practice guidelines for STS operations as identified by the International Maritime Organisation (IMO). The STS Operations Plan may be incorporated into an existing Safety Management System (SMS) required by Chapter IX of the *International Convention for the Safety of Life at Sea*, (SOLAS) 1974, as amended, if that requirement is applicable to the oil tanker in question. Any oil tanker subject to this chapter and engaged in STS operations shall comply with its STS Plan.

The purpose of the plan is to provide guidance to the Master and ship's officers directly involved in Ship-to-Ship Transfer (STS) operations with respect to the steps to be followed when this operation is likely to occur. Changes to the appendices generally do not require approval. The appendices should be maintained up to date by the vessel's managers, owners, and operators. Routine drills conducted on board will not only ensure that the ship's staff are familiar with their duties but will assist in forming a proficient team to combat all pollution incidents in an efficient manner. The purpose of this STS Operations Plan is to provide guidance to the master and or STS operator on board the ship in conducting the transfer in a safe and efficient manner. The STS operation are divided into five phases:

(1) Pre-arrival planning
(2) Arrival
(3) Berthing

© The Author(s), under exclusive license to Springer Nature Switzerland AG 2025

A. A. Olsen, *Introduction to Ship-to-Ship Transfers for Oil Tankers*, Synthesis Lectures on Ocean Systems Engineering, https://doi.org/10.1007/978-3-031-74802-8_2

(4) Cargo transfer; and

(5) Departure.

During each phase of the operation there are different procedures to follow and check-lists to complete. A risk assessment should be carried out before operations commence and a contingency plan should be put in place to deal with emergencies.

2.2   Scope

Each oil tanker involved in the cargo transfer operation should have on board a plan prescribing how to conduct STS transfer operations The Manual should be written in the working language of the master and officers and, if the working language of the master and officers is not English, French, or Spanish, include a translation into one of these languages. A copy of the STS Plan should be available at the following locations on each oil tanker:

- The navigation bridge
- The cargo transfer control station; and
- The engine-room.

Pre-arrival Planning

3

3.1 Conditions and Requirements

3.1.1 Ship Compatibility

It should be ensured that the ships are compatible in design and equipment; that they comply with the various recommendations included in this plan; and that mooring operations, hose handling and communications can be conducted safely and efficiently. It is essential that information relating to the overall dimensions, freeboard, position of manifolds, mooring points and fenders should be passed to the master(s) of the ship(s). It is recommended that ships with bridge wings that extend beyond the ship's maximum breadth are not used for STS transfer operations. The following should be determined prior to berthing (Table 3.1).

3.1.2 Person in Overall Advisory Control (POAC)

A ship-to-ship transfer operation should be under the advisory control of a designated Person in Overall Advisory Control (POAC). The POAC will either be one of the Masters of the vessels concerned or an STS Superintendent, Lightering Coordinator or Mooring Master employed by an STS Resource Provider. It is not intended that the POAC in any way relieves the ships' Masters of any of their duties, requirements, or responsibilities. The Flag State Administration, cargo owners or oil tanker's operators should agree and designate the POAC who should have at least the following qualifications:

© The Author(s), under exclusive license to Springer Nature Switzerland AG 2025   11
A. A. Olsen, *Introduction to Ship-to-Ship Transfers for Oil Tankers*, Synthesis Lectures
on Ocean Systems Engineering, https://doi.org/10.1007/978-3-031-74802-8_3

Table 3.1 Cargo handling compatibility

Cargo handling compatibility	
(1)	The size and number of manifolds to be used
(2)	The minimum and maximum expected height of the manifold above the waterline during the transfer operation, and the freeboard differences during the cargo transfer
(3)	Whether the cargo cranes or derricks are in satisfactory condition and of suitable safe working load (SWL)
(4)	The hose supports at the ship's side are the adequate to prevent damage to the hose through chafing
(5)	That both ships have manifolds that comply with *OCIMF Recommendations for Oil Tanker Manifolds and Associated Equipment*

- An appropriate management level deck license or certificate meeting international certification standards, with all STCW and dangerous cargo endorsements up to date and appropriate for the ships engaged in the STS operation
- Attendance at suitable ship-handling course
- Conduct of a suitable number of mooring/unmooring operations in similar circumstances and with similar vessels
- Experience in oil tanker cargo loading and unloading
- A thorough knowledge of the geographic transfer area and surrounding areas
- Knowledge of spill clean-up techniques, including familiarity with the equipment and resources available in the STS contingency plan; and
- Thorough knowledge of the STS Plan.

The POAC should:

- Ensure that the cargo transfer, mooring, and unmooring operations are conducted in accordance with the required STS plan, the contents of this chapter of the Manual and take into account the recommendations contained in the industry publication *"Ship to Ship Transfer Guide – Petroleum"*
- Advise the Master(s) of the critical phases of the cargo transfer, mooring and unmooring operation
- Ensure the provisions of the contingency plan are carried out in the event of a spill
- Ensure that all required reports are made to the appropriate authorities
- Ensure that crewmembers involved in each aspect of the operation are properly briefed and understand their responsibilities
- Ensure that approach and mooring operations are not attempted until proper effective communication has been confirmed between the two oil tankers and appropriate checks have been completed

- Ensure that a pre-transfer STS safety check is undertaken in accordance with accepted industry guidance; and
- Ensure that appropriate checks are undertaken prior to unmooring.

The POAC should have the authority to advise:

- Suspend or terminate the transfer operation; and
- Amend the transfer plan for the particular operation.

3.1.3 Recording and Checklists

The STS operation should be recorded in the Oil Record Book Part II, as required for all oil cargo transfer and ballast operations. The checklists contained in Appendix A.1 of the *Ship-to-Ship Transfer Guide—Petroleum (ICS)*, should be followed and filled out concurrently with the STS operation.

All records should be retained onboard for at least three years

3.1.4 Approval from Authorities

When the STS transfer is performed within the territorial waters or Exclusive Economic Zone (EEZ) of a country, local and national regulations should be checked, and appropriate approvals obtained. The responsible person should inform the appropriate authorities of STS transfer operations to be conducted in the lightering area (Table 3.2).

Where, in an exceptional case, all the information specified above is not available more than 48 h before the STS operations are to take place, the oil tanker discharging the oil cargo shall notify the authorities not less than 48 h in advance that an STS operation will occur. The information specified above shall be provided to the authorities at the earliest opportunity. Once the initial report for any ETA has been made, it should, if possible, be updated when a variance of more than two hours is expected from the time given in the latest report.

Table 3.2 Advance notice to authorities

Notice to authorities minimum 48 h in advance	
(1)	Name, Flag, call sign, IMO Number and estimated time of arrival of the oil tankers involved in the STS operations
(2)	Date, time, and geographical location at the commencement of the planned STS operations
(3)	Whether STS operations are to be conducted at anchor or underway
(4)	Oil type and quantity
(5)	Planned duration of the STS operations
(6)	Identification of STS operations service provider or person in overall advisory control and contact information
(7)	Confirmation that the oil tanker has onboard an STS Operations Plan meeting the requirements of MARPOL Annex I, Chap. 8, Regulation 41, 42, and 43

3.1.5 Transfer Area

The STS transfer area should be specially selected for safe operations, in coordination with appropriate authorities. Points to be considered when selecting the transfer area (Table 3.3).

3.1.6 Weather Conditions

The weather condition limit will mostly depend on the effect of the sea and swell on the fenders or mooring lines and the rolling movements induced by the participating ships, taking into account the relative freeboard and displacement. The following items should be taken into account regarding weather condition:

- If the transfer takes place with one ship at anchor, special care should be given to the ultimate strain placed on the anchor cable due to yawing movements aggravated by the current and weather condition
- Weather forecasts should be obtained before and during the transfer
- Throughout the berthing operation the visibility should be good enough for safe manoeuvring, taking into account the safe navigation and collision avoidance requirements; and
- Special attention should be given to electrical storms.

Table 3.3 Transfer area

Transfer area	
(1)	Notify and obtain agreement from the applicable coastal authority
(2)	The shelter provided from the weather, particularly from sea and swell
(3)	Present and forecast weather conditions
(4)	Tidal current conditions
(5)	Safe distances from offshore installations
(6)	The availability of a designated lightering area
(7)	Sufficient sea-room to be available to allow for normal drift or streaming patterns when the cargo transfer operations are conducted underway
(8)	Sufficient sea-room and water depth required for manoeuvring during berthing and unberthing
(9)	Locations of underwater pipelines, cables, artificial reefs, or historic sites
(10)	Selection of a safe anchorage with sufficiently good holding ground
(11)	Traffic density
(12)	Availability of emergency and oil spill response capability
(13)	Distance from shore logistical support
(14)	Security threats

3.2 Communications

3.2.1 Language

A common language for communication should be agreed before operations commence.

3.2.2 STS Instructions

The organisers generally provide STS instructions. This may be the operator of the ships if carrying out "in-house" operations or it may be an STS Service provider. Normally such providers send advance STS instructions to the ships concerned. The following information should be sent to the ship by the STS organisers (Table 3.4).

3.2.3 Navigational Warnings

The following navigational warnings should be broadcasted, before and during the transfer operation, to all ships advising (Table 3.5).

Table 3.4 Information to be sent to the ship by the STS organiser

Information to be sent to the ship by the STS organiser	
(1)	The organiser's full title, identification of person in overall advisory control and contact numbers
(2)	Description of the planned STS operation including the location of the transfer area
(3)	Details of equipment (including confirmation of integrity of the fenders, hoses, etc.), logistical support and personnel to be provided
(4)	Requirements for the preparation of moorings, manifolds and lifting gear
(5)	Local and national STS regulations, where applicable
(6)	Identity of the STS service provider and/or STS Superintendent

Table 3.5 Navigational warnings during transfer

Navigational warnings	
(1)	The name and flag of the ships involved
(2)	Geographical position of operations and general headings
(3)	Nature of operations
(4)	Time of starting operations and expected duration
(5)	Request for wide berth and the need to exercise caution when navigating in the transfer area

3.2.4 Communications During Operations

Contact should be established on the appropriate VHF channel at the earliest opportunity. Approach, mooring and unmooring should not be attempted until proper effective communication has been confirmed between the two ships. It should be confirmed that the portable radios on each ship are capable of working on the same frequencies, alternatively that one of the vessels have enough portable devices to adequately supply both vessels. The ship's officers responsible for mooring stations should be provided with portable radios. Ship's emergency portable VHF radios should not be used for routine operations.

3.2.5 Procedures for Communication Failure

If communication breakdown occurs during the approach manoeuvre, the manoeuvre should be aborted if appropriate and safe to do so and the subsequent actions taken by each ship should be indicated by the appropriate sound signals as prescribed in the *International Regulations for Preventing Collisions at Sea (COLREGS)*. If communication is lost during cargo transfer the emergency signal should be sounded and all operations in progress should be suspended immediately if it is safe to do so. Operations should not be resumed until satisfactory communications have been re-established.

3.3 Equipment

3.3.1 Fenders

Fenders may be secured in place on either ship, but it is recommended that they are placed on the manoeuvring ship. It should be noted that where fenders are to be rigged on the manoeuvring ship there may be more stresses on the head fender wire, and that the fender wire utilises one head wire per winch (Figs. 3.1 and 3.2).

When fenders are fitted to the manoeuvring ship, primary fenders should be positioned one at each end of the parallel body, with similar additional units fitted in between (refer to Fig. 3.3). The fender string may be made up to a pre- arranged length. Alternatively in some operations where four fenders are used, it has been found beneficial to position them in two groups of two. In this way, and with each group positioned well forward or well aft on the parallel body, better protection can be provided. Secondary fenders may be positioned fore and aft of the parallel body. Fender moorings should be monitored frequently and tended as necessary to ensure that they do not become too slack or too taut and that the fenders remain in position. The length of the fender string should be such that the fenders will be able to distribute the maximum anticipated impact load within the parallel body of both ships.

Fig. 3.1 Fenders rigged in continuous string

Fig. 3.2 Fenders rigged in pairs

Fig. 3.3 Tankers lightering whilst tethered

If fenders are provided by an STS Service provider, the ship's Master, responsible person, or organiser should ascertain the age of the fenders to be used and should be satisfied that reasonable measures have been taken to ensure that they are fit for the intended service. The fender certificates should be made available to assist with this (Table 3.6).

3.3.2   Hoses

The diameter for a chosen cargo transfer hose is governed mainly by the required flow rate and the manifold dimensions. Hose lengths should be considered on a case-by-case bases and but as a reference, hose lengths equal to twice the maximum difference in manifold height between the two ships are usually sufficient to allow for variables during transfer. A rule of thumb for calculating the minimum bending radius (MBR) of a rubber hose is given by the formula:

$$MBR = No\min al\,Bore\,of\,Hose \times 6$$

As the tanker rises or falls as a result of cargo transfer, the hose should be adjusted so as to avoid undue strain on the hoses, connections and ship's manifold and to ensure that the radius of the curvature of the hose remains within the limits recommended by the manufacturer. A visual inspection of each of the hose assemblies should be carried out before they are connected to the manifolds to determine if any damage has been caused. If hoses are provided by an STS Service provider, the Master, shipping company or organiser should ascertain the age of the hoses to be used and should be satisfied that reasonable measures have been taken to ensure that they are fit for the intended service. The hose certificates should be made available by the STS Service provider.

3.3.3   Hose Handling

Although oil hoses are robustly constructed for a marine environment, they can be damaged from improper handling. In general, while handling hoses, adequate support is the key to the prevention of over-bending (kinking), which can lead to premature hose retirement. When transferring the one end of the hose to the other ship, lifting straps should be used that are preferably flat nylon or equivalent reinforced cloth bands, and at least 150 mm (5.9 in) wide to prevent any chafing of the hose cover. If nylon or equivalent straps are not available, the best substitute is a sling of large circumference nylon or polypropylene rope. Wire should not be used.

Table 3.6 Fender selection assistance request form

FENDER SELECTION ASSISTANCE REQUEST FORM		
For Ship-to-Ship Use		
(To be filled out prior to contacting fender providers)		

Location of site: _____

Potential sea state: _____   Potential Beaufort Scale: _____

	Ship A	Ship B
Type of ship		
Displacement Tonnage (start of STS ops.) (DT)		
Gross tonnage (GT)		
Deadweight tonnage (DWT)		
Length overall (LOA)		
Loaded draught		
Beam		
Freeboard when coming into contact		
Relative approach velocity of ships		

Other relevant information

3.3.4 Pipe Reducers

It should be ascertained that adequately sized pipe reducers with packings are available onboard, to accommodate actual manifold and hose dimensions.

3.3.5 Mooring Equipment

Typically, a mooring pattern for exposed locations for lightering vessels not fitted with special mooring arrangements would consist of at least six headlines, two forward and two back springs, and four stern lines. Where specialised mooring equipment is fitted (e.g., on dedicated STS transfer ships) the number of headlines could be reduced to four where this has proven to be reliable for the local operating environment.

3.3.6 Personnel Transfers

Transfer of personnel between ships should be kept to an absolute minimum.

3.3.7 Lighting

During STS transfers at night, normal in-port deck lighting should be adequate. The minimum recommended lighting is five foot-candles (lumens) at transfer connection points and one footcandle in oil transfer operation work areas (measured 1 m (3.2 ft) above the deck). Portable spotlights, which should be flameproof, and bridge wing spotlights are useful for night mooring and unmooring operations.

3.3.8 Ancillary Equipment

All ancillary equipment—wires, messengers, stoppers, strops, and shackles etc. should be inspected for condition prior to commencing the STS operation.

Table 3.7 Emergency situations

Actions to be taken during emergency	
(1)	Stop the transfer
(2)	Sound the emergency signal
(3)	Inform both crews on the ships of the nature of the emergency
(4)	Position all emergency stations
(5)	Implement emergency procedures
(6)	Drain and disconnect cargo hoses
(7)	Send mooring gangs to stations
(8)	Confirm the ships main engine is ready for immediate use
(9)	Advise standby boat of the situation and any requirements

3.4 Emergency

3.4.1 Emergency Signal

The agreed signal to be used in the event of an emergency on either ship should be clearly understood by the personnel on both ships. An emergency on either ship should be indicated immediately by sounding the ship's internal alarm signal and by sounding seven or more short blasts on the whistle to warn the other ship. All personnel should then proceed as indicated by the contingency plan.

3.4.2 Emergency Situations

In an emergency, the Master(s) involved should assess the situation and act accordingly. The following actions should be taken, or considered, in the event of any emergency arising (Table 3.7).

In addition, the ship's Master should decide jointly, particularly in the case of fire, whether it is to their mutual advantage for the ships to remain alongside each other. The basic actions, as listed above, should be included in individual STS contingency plan and be consistent with the ship's Safety Management System.

3.4.3 Emergencies During Manoeuvring

The Masters of both ships and the POAC should always be prepared to abort a berthing operation if necessary. The decision should be taken in ample time while the situation is still under control. The Masters of both ships should be immediately informed of each other's actions. The *International Regulations for Preventing Collisions at Sea (COLREGS)* must be complied with.

3.4.4 Procedures in the Event of Gas Accumulation on Deck

The transfer operation should be suspended if excessive cargo vapours are detected around the decks or manifolds of either ship and should not be resumed until the risk to both ships and their personnel is considered to have been averted.

3.4.5 Accidental Cargo Release

Any leakage or spillage should be reported immediately to the officers on cargo watch who should stop the cargo transfer and advise the person in overall advisory control. The transfer must remain suspended until it is agreed between the relevant persons/authorities that it is safe to resume.

3.4.6 SOPEP or VRP

Risk of oil pollution during STS transfer operations need not be greater than during in-port cargo transfers. However, as a transfer area may be out of range of port services, a contingency plan within the Shipboard Oil Pollution Emergency Plan (SOPEP) or Vessel Response Plan (VRP) to cover such risk should be available and should be activated in the case of an oil spill.

3.4.7 State of Readiness for an Emergency

The following arrangements should be made on both ships (Table 3.8).

3.5 Risk Assessment

A risk assessment should be carried out before each STS transfer operation and should cover operational hazards and the means by which they are managed. As a minimum, the risk assessment should (Tables 3.9 and 3.10).

The scope of the Risk Assessment should include confirmation of the following.

The level of complexity required will depend on the type of operations. For a particular transfer area utilising standard approved STS equipment and ships that are fully

Table 3.8 State of readiness for emergency situations

Prepared for emergency	
(1)	Main engines and steering gear ready for immediate use
(2)	Cargo pump and all other equipment trips relevant to the transfer to be tested prior to the operation
(3)	Crew available and systems prepared to drain and disconnect hoses at short notice
(4)	Oil spill containment equipment prepared and ready for use
(5)	Mooring equipment ready for immediate use and extra mooring lines ready at mooring stations as replacements in case of breakage
(6)	Firefighting equipment ready for immediate use

Table 3.9 Risk assessment guidelines

Risk assessment guidelines	
(1)	Identify the hazards associated with the operation (collision risks in the vicinity, cargo vapour pressure, H_2S content, etc.)
(2)	Assess the risks according to the probability and consequence
(3)	Identify the means by which to prevent and/ or mitigate the hazard
(4)	Contain procedures for dealing with unanticipated events

Table 3.10 Risk assessment scope

Risk assessment scope	
(1)	Adequate training, preparation, or qualification of oil tanker's personnel
(2)	Suitable preparation of oil tankers for operations and sufficient control over the oil tankers during operations
(3)	Proper understanding of signals or commands
(4)	Adequate number of crew assigned to controlling and performing oil transfer operations
(5)	Suitability of the agreed STS plan
(6)	Adequate communications between oil tankers or responsible person(s)
(7)	Proper attention given to the differences in freeboard or the listing of the oil tankers when transferring cargo
(8)	The condition of transfer hoses
(9)	Methods of securely connecting hose(s) to the oil tanker(s) manifold(s)
(10)	Recognition of the need to discontinue oil transfer when sea and weather conditions deteriorate
(11)	Adequacy of navigational processes

operational, a generic risk assessment might be appropriate. For STS operations being undertaken in a new area, or in the event of a deviation from a routine STS transfer, e.g., VLCC to VLCC operation, a risk assessment should be carried out for each 'non-standard' activity and the Master shall advise the Company. Operations shall not proceed until a full risk assessment has been carried out and agreed by the Master and the Company.

3.6 Contingency Plan

A contingency plan should be put in place covering all possible emergencies, especially during the manoeuvring and cargo transfer phase, and provide for a comprehensive response. In addition, contingency plans should have relevance to the location of the operation and take into account the resources available both at the transfer area and with regard to nearby back-up support. Where appropriate, the contingency plan should be integrated with similar plans prepared by the responsible local authority. The contingency plan should be agreed between both ships, the STS organiser and the local or national authorities (as appropriate) before STS operations commence. The lightering/receiving ship will generally be playing the lead role in an STS transfer operation. Accordingly, where organisers have delegated the preparation of a contingency plan, it will normally be incumbent on the Master of such a ship to establish the overall plan that will be reviewed and agreed.

The SOPEP or VRP identifies measures to deal with operational oil spills and also spills resulting from casualties. Please see the appropriate section of the SOPEP/SMPEP or VRP for measures to be implemented during such emergencies. The following Emergency Procedures should be taken into account when preparing the contingency plan:

(1) Each vessel must have emergency duties assigned to designated members of the crew in case of accidents that may arise during the transfer of oil, particularly in the case of spillages of oil. Refer to the SOPEP or VRP for list of responsible persons and actions regarding oil spills

(2) Having discovered a spillage, the operation should be stopped, and the immediate measures set forth in the contingency plan should be implemented. The appropriate authorities should be informed of any oil spillage together with size, nature, and cause. Each case of oil spillage must be entered in the Oil Record Book; and

(3) In case of spillage of 100 tons and above a report is to be prepared in the form recommended by the IMO. It should be forwarded to the Administration of the coastal State, or to the Flag State Administration if the vessel is in waters beyond the jurisdiction of the coastal state. The report should be in accordance with the *IMO Interim Guidelines for Reporting Incidents Involving Harmful Substances*.

3.7   Safety

3.7.1   General Safety

For all STS transfer operations each Master remains at all times responsible for the
safety of his own ship, its crew, cargo and equipment and should not permit safety to
be prejudiced by the actions of others. Each Master should ensure that the procedures
recommended by this STS transfer operations plan are followed and, in addition, that
internationally accepted safety standards are maintained. In this regard, the most promi-
nent international safety manual in use for cargo handling advice is the *International
Safety Guide for Oil Tankers and Terminals (ISGOTT)*.

3.7.2   Prevention of Fatigue

To prevent human fatigue during the STS transfer operation, the POAC and/or all the
responsible officers for the lightering operation should comply with rest period requirements of relevant ILO, IMO and national regulations, and the relevant section of the ISM.
Records of rest and work hour compliance should be retained.

3.7.3   Safe Watch Keeping

The Master should take into consideration the estimated duration of operations to ensure
that safe and fatigue-free watch keeping can be maintained throughout. For reasons of
crew workload, transfer operations taking place simultaneously from either side of the
STBL are generally not recommended unless fully reviewed by risk assessment.

3.7.4   Helicopter Operations

While ships are moored together, helicopter operations should not be permitted without the prior approval of the organisers, both ships' Masters, and responsible person. If
approved, the responsible person will co-ordinate the operations locally. No helicopter
operations are to be carried out during transfer of cargo and/or bunkers and/or ballasting
into cargo tanks.

Arrival at Port

4

4.1 Operational Procedures Before Manoeuvring

4.1.1 Preparation of Ships

See Tables 4.1 and 4.2.

The following information should be exchanged between the two ships.

4.1.2 Navigational Signals

The lights and shapes to be shown, and the sound signals made are those required by *the International Regulations for Preventing Collisions at Sea (COLREGS)*, and local regulations. These lights and shapes should be checked and rigged ready for display prior to the STS operation.

© The Author(s), under exclusive license to Springer Nature Switzerland AG 2025   27
A. A. Olsen, *Introduction to Ship-to-Ship Transfers for Oil Tankers*, Synthesis Lectures
on Ocean Systems Engineering, https://doi.org/10.1007/978-3-031-74802-8_4

Table 4.1   Preparation before manoeuvres begin

Preparation before manoeuvres begin	
(1)	Ensure that the crew is fully briefed on procedures and hazards, with particular reference to mooring and unmooring
(2)	Ensure that the oil tanker conforms to relevant guidelines, is upright and at a suitable trim
(3)	Confirm that all essential cargo and safety equipment has been tested
(4)	Confirm that mooring equipment is prepared in accordance with the mooring plan
(5)	Fenders and transfer hoses are correctly positioned, connected, and secured
(6)	Cargo manifolds and hose handling equipment are prepared
(7)	Obtain a weather forecast for the STS transfer area for the anticipated period of the operation
(8)	Agree the actions to be taken if the emergency signal on the oil tanker's whistle is sounded

Table 4.2   Information to be exchanged between ships

Preparation before manoeuvres begin	
(1)	Mooring Arrangements
(2)	Quantities and characteristics of the cargo(es) to be loaded (discharged) and identification of any toxic components
(3)	Sequence of loading (discharging) of tanks
(4)	Details of cargo transfer system, number of pumps and maximum permissible pressure
(5)	Rate of oil transfer during operations (initial, maximum, and topping-up)
(6)	The time required by the discharging oil tanker for starting, stopping, and changing rate of delivery during topping-off of tanks
(7)	Normal stopping and emergency shutdown procedures
(8)	Maximum draught and freeboard anticipated during operations
(9)	Disposition and quantity of ballast and slops and disposal if applicable
(10)	Details of proposed method of venting or inerting cargo tanks
(11)	Details of crude oil washing, if applicable
(12)	Emergency and oil spill containment procedures
(13)	Sequence of actions in case of spillage of oil
(14)	Identified critical stages of the operation
(15)	Watch or shift arrangements
(16)	Environmental and operational limits that would trigger suspension of the transfer operation and disconnection and unmooring of the tankers
(17)	Local or government rules that apply to the transfer
(18)	Coordination of plans for cargo hose connection, monitoring, draining and disconnection
(19)	Unmooring plan

Berthing

5

5.1 Manoeuvring

5.1.1 Basic Berthing Principles

Berthing and unberthing operations should be conducted during daylight unless the personnel concerned are suitably experienced in night-time STS manoeuvring operations. For some inshore areas the port authority may require a pilot to be taken. In such circumstances the pilot should advise on all aspects of navigation and piloting, but the Master remains in overall control and in command of his ship.

5.1.2 Manoeuvring Alongside with Two Ships Under Power

The larger of the two ships should maintain steerage way at slow speed (preferably about 5 knots (5.7 mph; 9.5 km/h) keeping a steady course heading. Local conditions and knowledge will dictate the appropriate heading with due regard to transfer area and weather conditions. The manoeuvring ship then manoeuvres alongside. It is recommended that the manoeuvring ship approaches and berths with the port side to the starboard side of the constant heading ship.

5.1.3 General Advice for Controlling the Two Ships

Each ship should take the following into account:

© The Author(s), under exclusive license to Springer Nature Switzerland AG 2025   29
A. A. Olsen, *Introduction to Ship-to-Ship Transfers for Oil Tankers*, Synthesis Lectures on Ocean Systems Engineering, https://doi.org/10.1007/978-3-031-74802-8_5

- Courses requested by the manoeuvring ship must be followed by the constant heading ship
- Ship's speed should be controlled by adjusting engine revolutions (or propeller pitch)
- Any adjustment should be limited; for example, to plus or minus 5 rpm rather than using the relatively coarse engine room telegraph system. However normal full operating range must remain readily available
- For diesel engines, ascertain number of air starts available
- At night the deck should be adequately lit and, if possible, the ship's side and fenders should be lit by spotlights
- The side for mooring should be clear of all over side obstructions; permanent and otherwise
- The navigation lights and shapes appropriate to STS transfers should be displayed
- There should be effective radio communications between the bridge and mooring personnel and
- There should be effective communications between the Masters of each ship
- Vessel A: Constant Heading Ship—Constant Speed (about 5 knots)
- Vessel B: Manoeuvring Ship (Fig. 5.1).

Fig. 5.1 A possible final approach manoeuvre

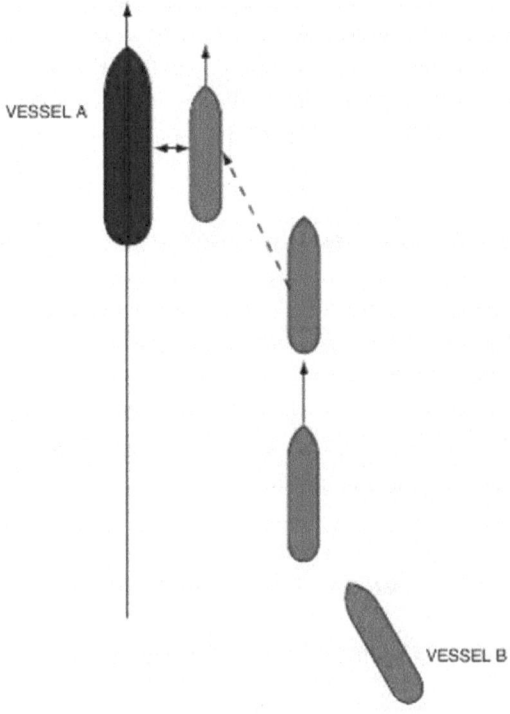

VESSEL A

VESSEL B

5.1.4   Advice for Manoeuvring Alongside

- If either of the Masters of the ships or the STS Superintendent has the slightest doubt about the safety of the manoeuvre, the berthing operation should be aborted
- At all times each ship is responsible for maintaining a proper lookout
- Generally, during manoeuvring, the wind and sea are kept on the port bow of the usually larger constant heading ship; however local conditions or knowledge may indicate an alternative approach
- The angle of approach adopted by the manoeuvring ship should not be excessive
- A common method of berthing is for the manoeuvring ship to approach the constant heading ship from the quarter on the side of berthing. On closer approach the manoeuvring ship should parallel the course of the constant heading ship at a safe distance that is appropriate to the conditions, before then positioning itself relative to the constant heading ship. Contact is made by manoeuvring ship reducing the distance by appropriate rudder and engine movements until the fenders touch
- The two ships should preferably make parallel contact at the same speed with no astern engine movements being necessary
- No engine movement on the constant heading ship should be made without advising the STS Superintendent or Master of the manoeuvring ship; and
- The effects of ship interaction should be anticipated when manoeuvring at close quarters.

5.1.5   Manoeuvring a Combined Two-Ship System to Anchor

On completion of mooring, the constant heading ship will usually power all future manoeuvres and, if a transfer at anchor is planned, will proceed to the agreed anchoring position. During this time, the (former) manoeuvring ship will have its engines stopped and rudder amidships. It should be emphasised that, for this period, in order to avoid problems for the manoeuvring ship the constant heading ship should not use strong astern engine movements. Speeds through the water should be minimal. The constant heading ship should use the anchor on the side opposite to that on which the other ship is moored. Once at anchor, each ship is responsible for watch keeping arrangements.

5.1.6   Underway Transfer

As long as adequate sea-room is available and traffic conditions, weather, sea conditions and forecasts are suitable, then transfers of this type can be carried out, but it should be noted that speeds through the water should be minimal. The constant heading ship maintains steerage way at slow speed on a steady course and the (former) manoeuvring ship keeps its rudder amidships and remains (with engines stopped) as a towed ship. In order to minimise towing loads on the moorings, the constant heading ship should alter her engine revolutions sparingly, adjusting speed very gradually. The chosen course and speed should be agreed by the two Masters' and the STS Superintendent and should result in minimum relative movement between the two ships and minimum turbulence in the gap between the hulls.

Under such circumstances, while the ships are moored together as a unit, safe navigation and collision avoidance is usually the responsibility of the constant heading ship but may be under the direction of the person in overall advisory control aboard the lightering ship. The two ships can drift freely provided conditions are suitable and a transfer area of suitable size is available. The use of the underway transfer system requires a full navigational watch to be kept on the bridge of each ship.

5.1.7 Manoeuvres with One Ship at Anchor

One ship anchors in a pre-determined position using the anchor on the side opposite to that on which the other ship will moor. The berthing operation should only commence after the anchoring ship is brought up to her anchor and is lying on a steady heading with reference to prevailing current and wind conditions. The Master of the ship which is to anchor should allow for the fact that the single anchor will be required to hold both ships.

When anchoring in deep water, and using an extended scope of cable, the Master of the ship that is to anchor should also ensure that the windlass is capable of recovering the cable and anchor once the operation is completed. The type of berthing operation then undertaken by the manoeuvring ship is similar to a normal approach to a jetty. A risk assessment should be undertaken by the organisers to evaluate the necessity of tug assistance for the manoeuvring ship. A careful watch should be kept on the heading of the anchored ship and the anchored ship should advise the manoeuvring ship immediately if she has any tendency to yaw. Where there is a tendency to yaw excessively, a tug should be employed to hold the anchored ship on a steady heading. If no tug is available, postponement of the operation should be considered. This manoeuvre can be preferred for more constrained transfer areas, especially when tug assistance is available, or if the manoeuvring ship is fitted with a bow thruster. Where current and wind are not from the same direction or the wind varies in speed or direction the anchored ship can yaw (or lie cross-current), making it difficult for the manoeuvring ship to berth alongside. Also, both ships could experience different effects due to their different freeboards and draughts. In these circumstances tug assistance may be advisable to hold the anchored ship on a steady heading during berthing. It is recommended that the services of an experienced STS Superintendent be utilised for this type of operation. However, berthing should not be attempted when the tidal stream is due to change.

When approaching a ship at anchor some master's recommend a wider angle of approach than that adopted for manoeuvres underway. A wider angle of approach, especially when tugs are not available, helps to avoid early ship-to-ship contact in cases where the anchored ship might yaw unexpectedly. It is recommended that the manoeuvring ship approaches and berths with her port side to the starboard side of the other ship. When mooring to an anchored ship, care should be taken not to pull the anchored ship quickly towards the manoeuvring ship.

Mooring

6

6.1 Mooring

6.1.1 Mooring Preparations

Moorings should be arranged and rigged to allow safe, effective line tending when the ships are secured together. This is very important on board the manoeuvring ship whose mooring lines will normally be used, but this should also be addressed on the constant heading ship where rope messengers have to be made ready between fairleads and deck winches. When preparing the mooring plan, the following factors should be taken into account (Table 6.1).

Most STS Service providers have a standard mooring plan, suitable for the particular location. It is important to ensure moorings allow for ship movement and freeboard changes to avoid over stressing the lines throughout the operation, but that they are not so long that they allow unacceptable movement between the ships. Mooring lines leading in the same direction should be of similar material. The order of passing mooring lines during mooring, and of releasing lines during unmooring, should be agreed. Where the STS Service providers utilise quick-release mooring hooks, their role and use should be discussed to ensure proper understanding. Figure 6.1 illustrates a typical and proven mooring plan for an STS transfer operation in offshore waters. Scope for additional head and stern lines is preferred, and at any time spare lines should be readily available to supplement moorings if necessary or in the event of a line failure.

Figure 6.2 is an example of a typical STS mooring arrangement.

The mooring lines are normally deployed from the manoeuvring ship. However, when prevailing weather conditions or weather forecasts require it, sending lines from both ships can increase the number of mooring lines. Loads should not be concentrated by passing most of the mooring ropes through the same fairlead or onto the same mooring bitts.

A. A. Olsen, *Introduction to Ship-to-Ship Transfers for Oil Tankers*, Synthesis Lectures on Ocean Systems Engineering, https://doi.org/10.1007/978-3-031-74802-8_6

Table 6.1 Mooring plan preparation

Mooring plan preparation	
(1)	Size of each ship and the difference in their sizes
(2)	The expected difference in freeboards and displacements
(3)	The anticipated weather and sea conditions
(4)	The degree of shelter offered by the location
(5)	The efficiency of the mooring line leads available

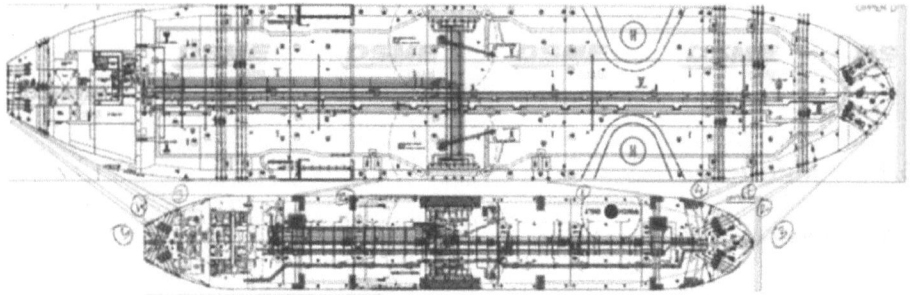

Fig. 6.1 Typical and proven mooring plan for an STS transfer operation in offshore waters

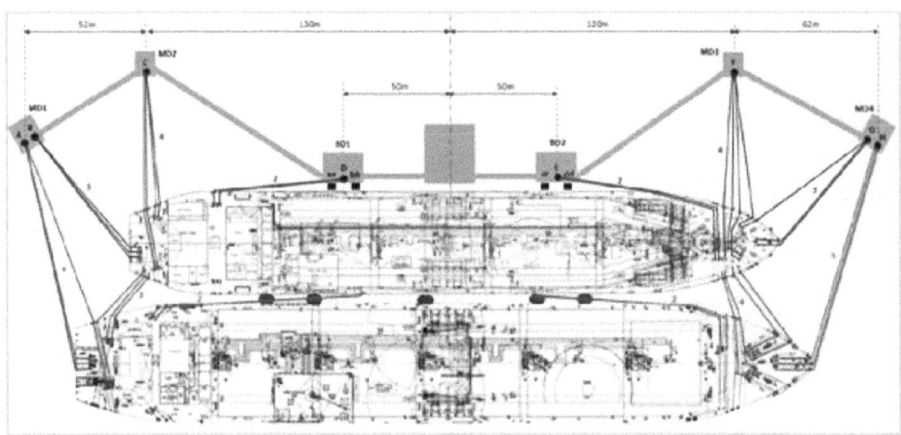

Fig. 6.2 Typical STS mooring arrangement

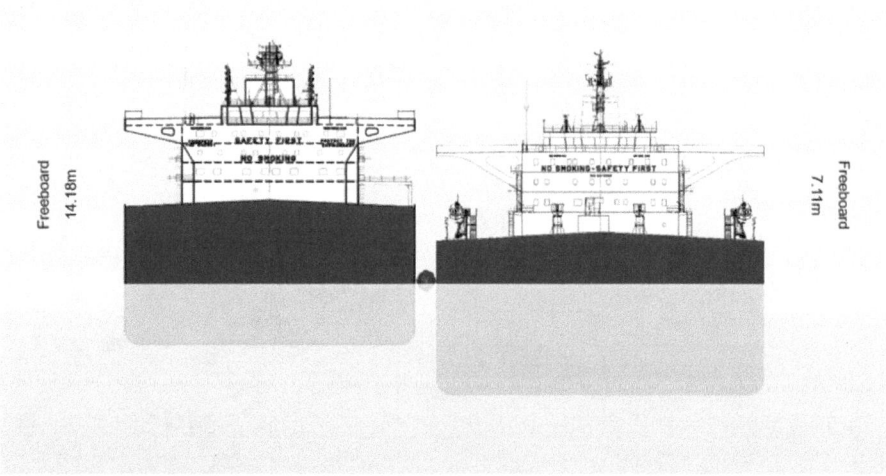

Fig. 6.3 Maximum and minimum freeboard heights

Use should be made of all available fairleads and bitts. During the operation, freeboard differences should be kept to a minimum, with consideration given to ballasting the higher ship and de-ballasting the lower one where this is possible. The steeper the orientation of the mooring lines, the less effective they will be in resisting horizontal loads. The maximum anticipated freeboard difference should therefore be taken into account when planning the mooring plan in order to ensure that the vertical angle of each mooring line stays as small as practicable throughout the operation. Refer to Fig. 6.3 for maximum and minimum freeboard heights.

A ship's standard complement of mooring lines is generally suitable for STS transfer operations, but ships equipped with steel wire or high modulus synthetic fibre mooring lines should fit soft rope tails to them. The connection between the primary line and the soft rope tail should be made with an approved fitting, e.g., Mandel or Tønsberg Shackle. Rope tails should be at least 11 m (36 ft) long and have a dry breaking strength at least 25% greater than that of the wires to which they are attached in accordance with *OCIMF Mooring Equipment Guidelines*. Soft rope tails fitted to wire moorings also introduce the benefit of making the cutting of mooring lines easier in an emergency and, for this purpose, long-handled firemen's axes or other suitable cutting equipment should be available at all mooring stations. Strong rope messengers should be readied on both ships and in addition rope stoppers should be rigged in way of relevant mooring bitts. Where possible, heaving lines and rope messengers should be made of buoyant materials. A minimum of four messengers should be provided and ready for immediate use. Non-pyrotechnic line-throwing equipment may be used to make the first connection (Fig. 6.4).

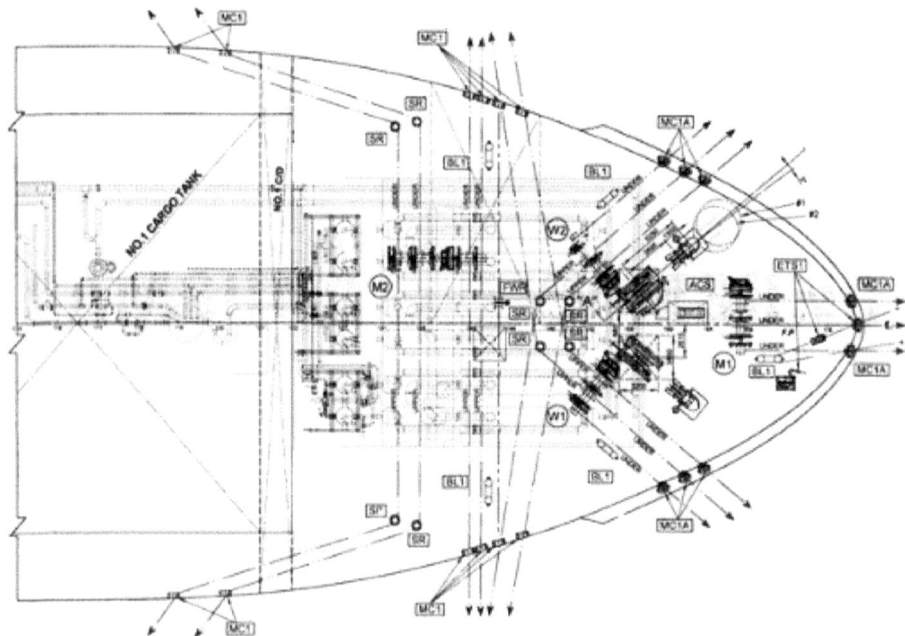

Fig. 6.4 Ship's standard complement of mooring lines

6.1.2   Tension in Mooring Lines

Excessive or uneven tension in mooring lines should be avoided because it can significantly reduce the weather threshold at which the forces in mooring lines will exceed their SWL. Attention should be given to this throughout the STS operation in order to ensure changes to the relative freeboards do not create excessive strain in the moorings.

6.1.3   Mooring Line Lead Angles

Peak loads on individual head and stern mooring lines can be minimised if the lead angles are similar and thus more effectively share the mooring loads.

6.1.4   Weather Thresholds

Higher weather thresholds for mooring loads can be tolerated when the STBL is at or close to fully loaded displacement. Masters' and persons in overall advisory control should be aware that weather thresholds may change significantly during the course of an STS operation as the STBL is lightened.

6.1.5 Long Period Waves

STS operations in locations subject to long period waves should be undertaken with caution. The load on mooring lines at any specific significant wave height greatly increases as the wave period, or period of encounter, increases.

6.1.6 Direction of Wave Encounter

Wave encounter from a beam direction during STS operations should be avoided. This will need particular attention when using exposed anchorage STS locations subject to strong currents, where the ships can lie at a large angle to the wind and waves. When conducting underway STS operations, the optimum wave encounter direction to control mooring loads is normally considered to be on the port bow, with the STBL to windward. But, depending on the relative size and displacements of the two ships, this may not always be the case. For instance, when both ships are of similar size and as the lightering ship's displacement increases relative to the STBL, the optimum mooring load wave encounter direction may change to the starboard bow, with the lightering ship to windward. In such a case, it may be advantageous to have sea-room available for course alterations across the wind (Fig. 6.5).

Fig. 6.5 Shuttle tanker Loch Rannoch discharging to the tanker British Hazel while alongside Jetty 4 at Sullom Voe Oil Terminal, Scotland

Cargo Transfer

7

7.1 Cargo Transfer

7.1.1 Pre-transfer Procedures

When the two ships are securely moored and before cargo transfer commences, good communications should be established between the personnel responsible for cargo operations on each ship and the pre-transfer checks should be satisfactorily completed. Before commencing the transfer operation, the responsible person(s) on the ships should ensure (Table 7.1).

7.1.2 Responsibility for Cargo Operations

Cargo transfer operations should be carried out in accordance with the requirements of the receiving ship. The person in charge of the cargo operations for each ship should be positively identified on a list posted on the bridge and cargo control room of both ships, together with the names of other persons supervising the cargo transfer. The transfer operation may be started only after the responsible person(s) on both oil tankers and the POAC have agreed to do so, either verbally or in writing.

7.1.3 Planning for Cargo Transfer

When preparing cargo loading and discharging plans, due regard should be given to ensuring that adequate stability is maintained, hull stresses remain within design limits and free surface effects are kept to a minimum throughout. The cargo transfer operation should

© The Author(s), under exclusive license to Springer Nature Switzerland AG 2025   41
A. A. Olsen, *Introduction to Ship-to-Ship Transfers for Oil Tankers*, Synthesis Lectures on Ocean Systems Engineering, https://doi.org/10.1007/978-3-031-74802-8_7

Table 7.1 Pre-transfer requirements

Mooring plan preparation	
(1)	Proper mooring of the ships
(2)	Availability of reliable communication between the two oil tankers
(3)	Emergency signals and shutdown signals are agreed
(4)	Proper connection and securing of hoses to the oil tanker's manifolds
(5)	Proper condition and position of hoses, hose saddles and supports
(6)	Flanged joints, where used, are fully bolted, and sealed and ensured oil tight
(7)	Proper blanking of unused cargo and bunker connections
(8)	Tools required for the rapid disconnection of hoses are located at the Manifold
(9)	Any valve through which oil could be discharged to the sea is closed and inspected and, if not used in the operation, is sealed to ensure that it is not inadvertently opened
(10)	Deck scuppers are properly plugged
(11)	Availability of empty drip trays on both oil tankers under couplings of hoses and means for drip tray drainage
(12)	Availability of materials on the oil tankers for on-deck clean-up in case of Spillage
(13)	Fire axes or suitable cutting equipment is in position at fore and aft mooring Stations
(14)	An engine-room watch will be maintained throughout the transfer and the main engine will be ready for immediate use
(15)	A bridge watch and/or an anchor watch will be established
(16)	Officers in charge of the cargo transfer are identified and details are posted
(17)	A deck watch is established to pay particular attention to moorings, fenders, hoses, and manifold integrity
(18)	Correct understanding of commands and signals by the responsible person(s) on the oil tankers during operations
(19)	Confirm completion of STS checklists

be planned and agreed in writing between the two ships and, where applicable, should include information on the following (Table 7.2).

Before starting cargo transfer, the discharging ship must be informed by the receiving ship of the flow rates required for the different phases of the cargo operation. If variations in transfer rate become necessary, the receiving ship should advise the discharging ship of its requirements. The discharging ship should similarly inform the receiving ship of any variations in flow rates due to its operations. The agreed transfer rate should not exceed the manufacturer's recommended flow rates for the cargo hoses.

Table 7.2 Cargo transfer operation information

	Cargo transfer operation information
(1)	Quantity of each grade of cargo to be transferred
(2)	Sequence of grades, cargo density, temperature, and specific precautions such as those that might be necessary for static accumulator products
(3)	Details of cargo transfer system, number of pumps, maximum pressure
(4)	Crude oil washing procedures
(5)	Initial and maximum topping off rates and notice period of rate change
(6)	Normal stopping and emergency shutdown procedures
(7)	Emergency and spill containment procedures
(8)	Watch or shift arrangements
(9)	Critical stages of the operation
(10)	Local or government rules that apply to the transfer
(11)	Material Safety Data Sheet (MSDS) in respect of cargo to be transferred to ensure that the receiving ship is aware of particular properties of the cargo, e.g., high Hydrogen Sulphide (H^2S) content, special firefighting requirements etc
(12)	Receiving ship to provide details of previous cargo
(13)	Coordination of plans for cargo hose connection, monitoring, draining and disconnection

7.2   Cargo Transfer

7.2.1   General

Throughout cargo transfer operations, the discharging ship and the receiving ship should station a responsible person at the cargo manifold area to observe the hoses and to check for leaks. In addition, throughout the cargo transfer, a responsible person equipped with a portable radio should be stationed at or near the cargo pump controls or in the cargo control room on the discharging ship, to take action as required. Cargo transfer should begin at the agreed slow rate to enable the receiving ship to check that the cargo pipeline system is correctly set. The transfer rate should also be reduced to an agreed topping off rate when the receiving ship's tanks are reaching their filling limits. Throughout the transfer, as a minimum hourly transfer rate checks and comparisons should be made between the two ships, and the results logged. Any differences or anomalies should be carefully checked and, if necessary, cargo operations should be suspended until the differences are resolved.

It should be ensured that adequate ullage space is left in each tank being filled. When it is required to stop cargo transfer operations, the responsible person should advise the pumping oil tanker in ample time. Cargo operations should be conducted under closed conditions, i.e., with ullage, sounding/sampling ports securely closed. In some parts of the world vapour balancing procedures are enforced and due regard should therefore be given to local regulations. Consideration should be given to the prevention of pressure surges by careful planning and control of pump speeds and the operation of valves. Static accumulator cargoes will require extra precautions and *ISGOTT* should be referred to when handling this type of cargo. During cargo transfer, appropriate ballast operations should be performed in order to minimise the differences in freeboard between the two ships, and to avoid excessive trim by the stern. Listing of either ship should be avoided, except as required for cargo tank draining on the discharging ship. Any national or local regulations controlling discharge of ships' ballast water should be complied with. Constant attention should be paid to mooring lines and fenders to avoid chafing and undue stress, particularly that caused by changes in relative freeboard. If at any time mooring lines need to be repositioned or adjusted this should only be done under strictly controlled conditions.

7.2.2   Suspension of Cargo Transfer

Both ships should be prepared to immediately discontinue the lightering operation, and to unmoor if necessary. The operation should be suspended when (Table 7.3).

Table 7.3 Suspension of cargo transfer

Suspension of cargo transfer	
(1)	Movement of the ships alongside reaches the maximum permissible and may cause loss of strength or excessive strain on hoses
(2)	Under adverse weather and/or sea conditions
(3)	Either vessel experiences a power failure
(4)	There is a failure of the main communication system between the ships and there are no proper standby communications
(5)	Escape of oil through sea valve or ship plating is discovered
(6)	There is an unexplained pressure drop in the cargo system
(7)	Fire danger is discovered
(8)	Any oil leakage is discovered from hoses, couplings, or the ship's deck piping
(9)	Any oil overflow on the ship(s) deck caused by over-filling of tank occurs
(10)	Faults or damage threatening the escape of oil are discovered
(11)	There is a significant, unexplained difference between the quantities of cargo delivered
(12)	There is any failure of the fendering system
(13)	When hours of work are exceeded on either ship

7.3   Vapour Balancing Considerations

Cargo operations should be conducted under closed conditions with ullage, sounding and sampling ports securely closed. Due regard must be given to any local regulations that may require the adoption of vapour balancing procedures.

7.3.1   Vapour Balancing Considerations Before Commencing Cargo Transfer

- Equipment should be provided on at least one of the vessels to enable the oxygen content of the vapour stream to be monitored. This should draw samples continuously from a location close to the vapour manifold connection and should include the facility for audible and visual alarms in the event that the oxygen content of the vapour stream exceeds 8% by volume. The oxygen analyser and associated alarms should be tested for proper function prior to each cargo transfer operation
- The oxygen content of the vapour space of each tank connected to the IG main in both ships should be checked and confirmed to be less than 8% by volume
- The vapour transfer hose should be purged of air and inerted prior to commencing transfer of vapours; and
- The vapour manifold valves should not be opened until the pressure in the cargo system of the receiving vessel exceeds that of the ship discharging cargo.

7.3.2   Vapour Balancing Considerations During Cargo Transfer

- The inert gas system on the discharging ship should be kept operational and on standby, with the inert gas main deck isolating valve closed. The inert gas system should be used if the inert gas pressure in the discharging vessel falls to a low level (300 mm (11.8 in) WG)
- The inert gas pressure on both ships should be monitored and each ship advised of the other's pressure on a regular basis
- No air should be allowed to enter the cargo tanks of the discharging ship
- Transfer operations should be suspended if the oxygen content of the vapour stream exceeds 8% by volume and should only be resumed once the oxygen content has been reduced to 8% or less by volume; and
- The cargo transfer rate must not exceed the design rate for the vapour balancing hose.

7.4   Safety During Cargo Transfer

The basic safety requirements for a transfer operation are similar to those for a normal port cargo operation as contained in the latest edition of *ISGOTT*. The following points are emphasised for an STS transfer operation:

7.4.1   Smoking and Naked Lights

Regulations regarding smoking and the use of naked lights should be strictly enforced. Warning notices should be displayed, and smoking rooms should be designated and clearly marked.

7.4.2   Earths on Electrical Switchboards

Earth indicator lights showing on the main switchboard indicate a faulty circuit and such faults should be immediately traced and isolated. This is to avoid the risk of arcing, especially in deck areas where hazardous accumulations of gas may be present.

7.4.3   Boilers and Diesel Engines

Precautions such as soot-blowing should be carried out prior to commencing the approach manoeuvre in order to avoid hot ash falling onto a ship's deck during the cargo transfer. In case of sparking from the funnel, transfer operations should be stopped immediately.

7.4.4   Ship-to-Ship Electric Currents

Elimination of electrical current and electrostatic charge in cargo hoses

In order to eliminate the potential for incentive arcing between the two ships when presenting the hose string for connection:

- A single insulating flange should be fitted within each hose string (or at one ship's manifold); or
- One length of electrically discontinuous hose should be fitted in each hose string; or
- Hoses that are specially constructed to prevent static build-up or electrical currents transferring between ships should be used.

The electrical potential between the two ships should be reduced to the minimum. Switching off cathodic protection systems of the impressed current type is not, in general, considered to be a feasible method of minimising ship-to-ship currents in the absence of an insulating flange or hose. If both ships have properly functioning impressed current cathodic protection systems, this is probably best achieved by leaving them running. Likewise, if one has an impressed system and the other a sacrificial system, the former should remain in operation. However, if either ship is without cathodic protection, or its impressed system has broken down, consideration should be given to switching off the impressed system on the other ship well before the two ships come together.

Other places where electrical arcing may occur
All ship-to-ship mooring lines should be insulated either by using the natural properties of soft mooring lines or by attaching a soft rope tail to the eye of each steel wire mooring line. If using soft rope tails, they should be of suitable length so that they extend to the outboard side of the ship receiving the mooring. Care should be taken to avoid low resistance ship-to-ship electrical contact in the following areas:

- Non-insulated metallic ladders or gangways between the ships—by the fitting of rubber ends
- Derrick or crane wire runners and hooks—by careful operation; and
- Unprotected bare wires and chains within fender support nets or cages—by good quality maintenance.

7.4.5 Use of Radio and Satellite Communications Equipment

Main radio equipment

Transmissions from the ship's main radio station can cause electrical resonance in insulated parts of some ship fittings, such as mast stays, and this can cause arcing across deck fittings. Similarly, arcing can occur on a ship's wireless aerials, especially over the surface of insulators when they have a coating of salt, dirt, or water. The use of the ship's main radio equipment during cargo operations can be dangerous. Radio transmissions should not be permitted during periods when there is the possibility of flammable gas in the region of the antennae or where there is doubt about the effective earthing of stays, derrick equipment and other such fittings. The main radio transmitting aerials on both ships should be earthed (grounded) and neither ship should use this equipment while alongside one another. Satellite communications equipment can be used for communications; however, the risks described below should be taken into account.

Satellite communications

Satellite communications equipment normally operates at 1.6 GHz and the power levels generated are considered to present few ignition hazards. However, this equipment should not be used when flammable gas is in the vicinity of antennae.

VHF and UHF radios

Any hand-held VHF and UHF radios, as used for mooring and cargo operations, should be of intrinsically safe manufacture.

Automatic identification systems (AIS)

Where either or both ships involved in STS operations are required to have an Automatic Identification System (AIS) operating while underway or at anchor, the AIS equipment should remain in use at all times, including during STS operations. The VHF equipment used for the AIS broadcasts need not be set to low power output during STS operations. However, during STS operations consideration should be given to using the optimal text entry area in the AIS message to include a phrase to indicate that the ship is restricted in her ability to manoeuvre, underway or at anchor, as a result of conducting STS transfer operations. It may be necessary to abbreviate the optional message to include this infor-mation. AIS broadcasts should not supplant the recommendation to broadcast navigational warnings by other means.

Portable electronic devices

It should be noted that portable cellular (mobile) telephones, pagers, cameras using bat-teries, Portable Data Assistants (PDAs), calculators etc., could constitute a risk to the ships if used in a hazardous area. Precautions should be taken to ensure that all personnel involved in the transfer, especially those who may be visiting the ships on other business (technicians, surveyors, etc.) are made fully aware of the dangers and any restrictions on the use of such items.

7.4.6   Use of Radar

General

The use of radar involves the operation of electrical equipment which is not intrinsically safe. Depending on the relative size of the two ships, during cargo transfer operations the radar beam of one ship may at times sweep the cargo deck of the other and be close enough to create potentially hazardous power densities in areas where flammable gas mixtures could be present. Consultation between master(s) is advisable before radar is used during cargo transfer operations. The following section gives further advice.

Use of 3 cm and 10 cm radar

Radiation from radar operating at frequencies above 9,000 MHz (3 cm) may be considered safe at distances of over 10 m. The radiated power from such radar should not present an ignition hazard provided scanners are correctly sited above the superstructure. Radar operating in the 3 cm waveband will normally be safe but should only be used with discretion. At the lower frequencies, as used by 10 cm radar, the possibility of induced arcing in parts of a ship's structure is present at ranges of up to 10 m (32 ft). Marine radars normally operate with a pulsed signal and a rotating scanner, so people are not continuously exposed to radiation. Therefore, the power-scanner interlocks should never be overridden without an appropriate risk assessment.

7.4.7 Gas Accumulation

The STS transfer operation should be suspended if cargo vapour accumulation around the decks or manifolds of either ship constitutes a risk to the ship or personnel and should not be resumed until it is considered safe to do so. The receiving ship should, prior to cargo transfer, provide details of the ship's previous cargo to the discharging ship (STBL). This will enable the discharging ship's personnel to take suitable precautions in the event that the previous cargo contained toxic vapours that could be displaced onto the deck of the discharging ship. Particular attention should be given to the potential of high H_2S levels in the cargo vapours and all necessary personal safety precautions should be taken.

7.4.8 Electrical Storms

When an electrical storm is present or imminent in the transfer area, the cargo transfer operation should be suspended and all vent risers, cargo systems and inert gas systems (IGS) secured until such time as it is considered safe to resume operations.

7.4.9 Galley Stoves

Before permitting the use of galley stoves and other cooking appliances while a ship is engaged in STS operations, the ship's master, and the STS Superintendent (if applicable) must, after taking into consideration the location, construction, and ventilation of the galley, jointly agree that no associated danger exists. Oil or gas fired stoves or electrical appliances using exposed elements should not be used.

7.4.10   Readiness for Firefighting Equipment

Fire-fighting equipment should be ready for immediate use on both ships. Foam monitors on each ship should be pointed towards the cargo manifold in use and left in a suitable condition for hands-off operation. Additional foam firefighting equipment should be immediately available for use on deck.

7.4.11   Accommodation Openings

All access doors to the accommodation should be kept closed during cargo transfer operations. The Master of each ship should designate those access doors that are to be used for personnel transit. Where possible, only doors remote from the main deck cargo area should be used. All doors opened for personnel transit should be closed immediately after use. The air conditioning intakes must be set to ensure the atmospheric pressure inside the accommodation is always greater than that of the external atmosphere. Air conditioning systems should not be set to 100% recirculation, as this will cause the pressure of the internal atmosphere to fall less than that of the external atmosphere, due to extraction fans operating in sanitary spaces and galleys.

7.4.12   Unauthorised Craft

No unauthorised craft should be allowed alongside either ship throughout the transfer.

7.5   Operations After Completion of Cargo Transfer

In accordance with previously agreed procedures, after completion of cargo transfer the following operations should be carried out (Table 7.4).

Table 7.4   Operations after completion of cargo transfer

Operations after cargo transfer	
(1)	All hoses should be drained into one ship prior to disconnecting. The oil tanker with the greatest freeboard should close the valve at the manifold and drain the oil contained in the hoses into the tank of the other oil tanker
(2)	Hoses should be disconnected and securely blanked
(3)	Cargo manifolds should be shut and securely blanked
(4)	Authorities should be informed of completion of cargo transfer and the anticipated time of unmooring

When the loading or discharging of cargo is completed, it must be recorded in the Oil Record Book (ORB).

Unmooring

8.1   Unmooring Procedure While One Ship is at Anchor

It is recommended that unmooring at anchor be carried out only by persons with considerable experience in STS operations and use of tugs should be considered where available, especially if yawing of the anchored ship is anticipated. It is also recommended not to unmoor during a change of tide. If, in the judgement of the POAC, the weather and current conditions so require, the constant heading ship should weigh her anchor and unmooring should be carried out while making way.

8.2   Unmooring Procedure After Underway Transfer

If the STS transfer operation have taken place while underway, it is normal to unmoor with the wind and sea on the port side and then bring the combined two-ship system head to the wind to spread apart the ships, unless local conditions dictate otherwise.

8.3   Unmooring Checks

Sufficient crew should be allocated to unmooring stations and consideration should be given to the following points (Table 8.1).

© The Author(s), under exclusive license to Springer Nature Switzerland AG 2025   53
A. A. Olsen, *Introduction to Ship-to-Ship Transfers for Oil Tankers*, Synthesis Lectures on Ocean Systems Engineering, https://doi.org/10.1007/978-3-031-74802-8_8

Table 8.1 Unmooring checks and considerations

Unmooring checks	
(1)	The cargo transfer side of the ship should be cleared of obstructions including derricks or cranes
(2)	The method of disengagement and letting go of mooring lines should be agreed
(3)	Fenders, including their towing and securing lines, should be checked to be in good order
(4)	Winches and winch lasses should be ready for immediate use
(5)	Rope messengers and rope stoppers should be ready at all mooring stations
(6)	Sharp fire axes or other suitable cutting equipment should be available at each mooring station
(7)	Communications should be confirmed between ships
(8)	Communications should be established with mooring personnel
(9)	Mooring personnel should be instructed to let go mooring lines only when directed
(10)	Shipping traffic in the vicinity should be checked
(11)	Checklist should be completed

9.1   Procedure for Unberthing

Special care needs to be taken when unmooring to avoid the two ships coming into contact. While there are other methods, a common method of unmooring is achieved by: The forward gang let go the forward Springs and then the Headlines. The after gang lets go the Stern lines and then the After Springs and the manoeuvring ship moves clear. Special care should be adopted during letting go of the last lines. This needs to be performed both safely and effectively. A method that can facilitate this is the "toggle pin technique." The mooring lines sequence may be changed at times according to weather conditions and/or at the discretion of the Mooring Master. The two vessels will be informed prior to the unmooring of any changes. The constant heading ship should not independently manoeuvre until advised that the manoeuvring ship is clear. It should be noted that local conditions or ship configurations may cause difficulties in separating the two ships and alternative plans should be considered.

9.1.1   Unmooring Using Quick Release Apparatus or Toggle Pins

Special care should be taken in regard to letting go the last lines in an expeditious and safe manner. This operation should be planned in advance, be undertaken by experienced crew and requires good communications and supervision. Different methodologies can be applied by STS Superintendents and ships' crews to carry out this task safely and effectively. One such method involves the use of quick release hooks secured around the mooring bitt or a "toggle" pin that is used in conjunction with a messenger to take the load of the mooring line while it is removed from the mooring bitt.

© The Author(s), under exclusive license to Springer Nature Switzerland AG 2025   55
A. A. Olsen, *Introduction to Ship-to-Ship Transfers for Oil Tankers*, Synthesis Lectures
on Ocean Systems Engineering, https://doi.org/10.1007/978-3-031-74802-8_9

Correction to: Introduction to Ship-to-Ship Transfers for Oil Tankers

Correction to:
A. A. Olsen, *Introduction to Ship-to-Ship Transfers for Oil Tankers*,
Synthesis Lectures on Ocean Systems Engineering,
https://doi.org/10.1007/978-3-031-74802-8

This book contains overlap in text with the previously published content [1, 2] that was inadvertently omitted. The authors failed to attribute or cite [1, 2]. The authors have now obtained permission to re-use this content from the American Bureau of Shipping and International Maritime Organization (IMO).

References

1. ABS rules and guidance, https://ww2.eagle.org/en/rules-and-resources/rules-and-guides.html.
2. IMO Publishing ePublications platform https://imo-epublications.org/ (digital),
    https://www.imo.org/en/publications/Pages/Distributors-default.aspx (print)

The updated version of this book can be found at
https://doi.org/10.1007/978-3-031-74802-8

© The Author(s), under exclusive license to Springer Nature Switzerland AG 2025   C1
A. A. Olsen, *Introduction to Ship-to-Ship Transfers for Oil Tankers*, Synthesis Lectures
on Ocean Systems Engineering, https://doi.org/10.1007/978-3-031-74802-8_10

Glossary of Terms

At Sea

The term "at sea" is used throughout this guide. It is intended to indicate offshore waters or partially sheltered waters. It may be, however, that an STS transfer operation at sea is to be conducted within the jurisdiction of a local (port) authority or national government. In such cases reference has to be made to local regulations and it may also be necessary to obtain local approval.

Closed Operations

Ballasting, loading, or discharging operations carried out without recourse to opening ullage and sighting ports. In these situations, ships will require the means to enable closed monitoring of tank contents, either by a fixed gauging system or by using portable equipment passed through a vapour lock.

Constant Heading Ship

During manoeuvring and mooring, the ship that maintains course and speed to allow the manoeuvring ship to approach and moor is referred to as the constant heading ship.

Dedicated Lightering Ship

A dedicated lightering ship is a ship designed to perform multiple STS operations. These ships are usually fitted with adequate primary and secondary (baby) fenders, which upon completion of an STS transfer are capable of being lifted and stowed in onboard cradles. They are usually fitted with their own hoses and are generally capable of performing STS operations without external assistance such as support craft. They may also be fitted with bow and stern thrusters, or large angle rudders to assist with manoeuvrability.

A. A. Olsen, *Introduction to Ship-to-Ship Transfers for Oil Tankers*, Synthesis Lectures
on Ocean Systems Engineering, https://doi.org/10.1007/978-3-031-74802-8

Discharging Ship

The ship containing cargo for transfer to the receiving ship, and which may also be known as the 'Ship to be Lightened' (STBL).

Manoeuvring Ship

During manoeuvring and mooring, the ship that approaches the Constant Heading Ship for mooring operations is referred to as the manoeuvring ship.

Organisers

Organisers are shore-based operators responsible for arranging an STS transfer operation. The Organiser may be an STS Service provider.

Person in Overall Advisory Control

The person agreed to be in overall control of an STS operation. It may be one of the Masters (generally the Master of the manoeuvring ship) or it may be an STS Superintendent.

Primary Fenders

Primary fenders are large fenders capable of absorbing the impact energy of berthing and wide enough to prevent contact between the ships should they roll while alongside one another.

Receiving Ship

The ship to which cargo is transferred from the Discharging Ship. The Receiving Ship may also be known as the lightering ship, or Service Ship (SS).

Secondary Fenders

Secondary (baby) fenders are fenders used to prevent contact between the two ships, should they be rolling or not parallel to each other. They are especially effective when rigged towards the ends of a ship and are of most benefit during mooring and unmooring operations.

Ship-to-Ship (Sts) Transfer Operation

An STS transfer operation is an operation where crude oil or petroleum products are transferred between seagoing ships moored alongside each other. Such operations may take place when one ship is at anchor or when both are underway. In general, the expression includes the approach manoeuvre, berthing, mooring, hose connecting, safe procedures for cargo transfer, hose disconnecting, unmooring and departure manoeuvre.

STS Service Provider

An STS Service provider is a company or organisation that specialises in providing services for the safe control of STS operations. The service provider may also supply the essential personnel and equipment needed such as hoses, fenders, and support craft.

STS Superintendent

A person who may be designated to assist a ships Master in the mooring and unmooring of the ships, and to co-ordinate and supervise the entire ship-to-ship transfer operation. He may also be known as Lightering Master or Mooring Master.

SWL

SWL or Safe Working Load is the operating limit to which equipment is tested for day-to-day use. Equipment should never be used beyond its SWL.

Transfer Area

A transfer area is an area within which an STS transfer operation takes place. Transfer areas should be selected in safe sea areas (refer to Sect. 3.1.5). In coastal areas they will be agreed with nearby coastal authorities and as appropriate, in accordance with specific port or national regulations. The expression "transfer at anchor" describes an operation where a cargo transfer is carried out

Transfer at Anchor

Transfer between ships when they are moored alongside each other and one of the ships is at anchor. The operation is an alternative to underway transfer.

Underway

By definition under the *International Regulations for Preventing Collisions at Sea* (COLREGS) a ship when underway is not at anchor, made fast to the shore or aground. However, she may be either steaming or drifting freely with current and weather.

Other Publications by the Same Author

Routledge and CRC Press

Olsen, Alexander. 2023. Merchant Ship Types. Routledge, London

Olsen, Alexander. 2023. Firefighting and Fire Safety Systems on Ships. Routledge, London

Olsen, Alexander. 2022. Core Principles of Maritime Navigation. Routledge, London

Olsen, Alexander. 2022. Introduction to Container Ship Operations and Onboard Safety. Routledge, London

Olsen, Alexander. 2023. Maritime Accident and Incident Investigation. Routledge, London

Olsen, Alexander. 2023. Maritime Cargo Operations. Routledge, London

Olsen, Alexander. 2023. Introduction to Ship Engine Room Systems. Routledge, London

Springer Series on Naval Architecture, Marine Engineering, Shipbuilding and Shipping (NAMESS)
Olsen, Alexander. 2024. Introduction to Celestial Navigation (Vol.15)

Olsen, Alexander. 2024. Introduction to Digital Navigation. (Vol.16)

Olsen, Alexander. 2024. Ship Operations in Extreme Low Temperature Environments. (Vol.19)

Olsen, Alexander. 2024. Safety Culture and Leading Indicators for Safety in the Maritime and Offshore Environment. (Vol.20)

Olsen, Alexander; Karkori, Fidaa. Containerized Cargo Handling and Stowage Principles and Procedures. (Vol.22)

Springer
Olsen, Alexander. 2024. Equipment Conditioning Monitoring and Techniques: Guidance for the Maritime Domain. Springer, Cham

Olsen, Alexander. 2024. Applying Physical Ergonomics to Modern Ship Design. Springer, Cham

Synthesis Lectures on Ocean Systems Engineering (SLOSE)
Olsen, Alexander. 2025. Dropped Object Prevention on Offshore Facilities and Installations Guidance for Safety Professionals and Practitioners.

Olsen, Alexander; Rossi-Ciampolini, Pamela. 2024. Ballast Water Treatment and Exchange for Ships.

Bibliography

Guide to Helicopter/Ship Operations (ICS)

Guidelines on the Use of High-Modulus Synthetic Fibre Ropes as Mooring Lines on Large Tankers (OCIMF)

International Chamber of Shipping and Oil Companies International Marine Forum, "Ship to Ship Transfer Guide (Petroleum)", Witherbys Publishing, Fourth Edition 2005.

International Chamber of Shipping, Oil Companies International Marine Forum and The International Association of Ports and Harbours, "International Safety Guide for Oil Tankers and Terminals", Witherbys Publishing and Seamanship International, Fifth Edition 2006.

International Convention on Standards and Training, Certification and Watch Keeping (STCW) Code (IMO)

International Maritime Organisation, "Manual on Oil Pollution, Section I, Prevention", Revised 1983

International Regulations for Preventing Collisions at Sea (COLREGS) (IMO)

Oil Companies International Marine Forum, "Guidelines for the Handling, Storage, Inspection and Testing of Hoses in the field," Witherbys Publishing, February 1995.

Oil Companies International Marine Forum, "Mooring Equipment Guideline," Witherbys Publishing, February 2007.

Oil Companies International Marine Forum, "Recommendations for Oil Tanker Manifolds and Associated Equipment," Witherbys Publishing, February 2007.

Recommendations for Ships' Fittings for Use with Tugs with Particular Reference to Escorting and Other High Load Operations (OCIMF)

Ship and Port Facility Security (ISPS) Code (IMO)

Standard Marine Communication Phrases (IMO)

A. A. Olsen, *Introduction to Ship-to-Ship Transfers for Oil Tankers*, Synthesis Lectures
on Ocean Systems Engineering, https://doi.org/10.1007/978-3-031-74802-8